A Christopher Ranch Story

Elephant Garlic Christmas

— Written by —

Ken Christopher

— Illustrated by —

Danny Voight

— Designed by —

Articulate Solutions, Inc.

This year, we shall be celebrating Christmas cheer, and it's with great pride that your attendance is requested. Sharing and caring bring joy to the land, and there's no better time than Christmas for us all to join hands. When the clock strikes at the four and the twelve let us all gather to celebrate this most festive of occasions. And while you all share your special bounty, there just may be a surprise for each and every one of you.

The King

Every year in the Kingdom of Gilroy all of our friends
gather to celebrate the most special of occasions.
Christmas comes but once a year, and it's the greatest
chance to share the season's cheer.

Dressed as an elf Mr. Mouse just can't wait,
to share all his cheese on this most important date.

MOWS9

rabbit

Mr. Rabbit's roasting carrots and he just can't wait,
but he mustn't take too long or he risks being late.

All dressed up in her Mrs. Clause design,
Mrs. Bear's homemade honey has finished just in time.

For a very special treat
Mr. Monkey knows what to do.
In bringing sweet bananas he'll
make their dreams come true.

When it comes to an elephant's Christmas,
there's one thing that's most nice.

Plenty of garlic, bringing
every dish the proper spice.

While our friends were celebrating, there was something just amiss, a new friend who's come to town who simply couldn't join the bliss.

Our newest Husky neighbor wants nothing more than to enjoy, but the sights and sounds and cheer that abounds, was much too much for him.

When the clock struck the four and the twelve, our young Elephant observed they were still by themselves.

"Whatever is wrong little Husky,
we're all gathered inside, together
we're safe there's no need to hide."

"I want to be there and join all the fun, but the lights and noise are too much for me."

"There's only one person who'll know
what to do, but he's late and he's truant,
let's find him, me and you."

Without missing a beat his ears were alerted, and in no time at all, "I found him," he blurted.

"Together we'll help him just you and I,
there's still time to save Christmas
before the night goes by."

"Oh Mr. Elephant you're just in time, my sleigh has been hobbled and I'm missing a ride."

"This is my moment, I know just what to do, Huskies are large and strong and I'll pull this sleigh."

On this Christmas night, our friends went dashing
through the snow, the King and the Elephant
impressed by just how fast he could go.

One by one the animals gathered
outside, the King, Elephant and
Husky all beaming with pride.

"You've done it Mr. Husky, you've done
your best, Christmas has been saved
thanks to your heart that's so brave."

"I've a solution tailored just for your cause."
  A brand new pair of earmuffs were placed in his paws.
"Our dear Mr. Husky can be overwhelmed by too much cheer,
  for him to join our Christmas, be mindful when he's near."

At last our friends gathered in delight, with garlicky dishes all around, it was the perfect Christmas sight.

"I know it wasn't easy, but you believed with all your heart, and from this special evening you'll never be apart."

At a quarter past nine, with the spirit of Christmas all around, our two friends slumbered away amidst their favorite food.

MAXWELL

THE END